Stamp More Stamps

by Miguel Delgado
illustrated by Paul Meisel

HMH

Copyright © by Houghton Mifflin Harcourt Publishing Company

All rights reserved. No part of this work may be reproduced or transmitted in any form or by any means, electronic or mechanical, including photocopying or recording, or by any information storage and retrieval system, without the prior written permission of the copyright owner unless such copying is expressly permitted by federal copyright law. Requests for permission to make copies of any part of the work should be submitted through our Permissions website at https://customercare.hmhco.com/contactus/Permissions.html or mailed to Houghton Mifflin Harcourt Publishing Company, Attn: Intellectual Property Licensing, 9400 Southpark Center Loop, Orlando, Florida 32819-8647.

Printed in the U.S.A.

ISBN 978-1-328-77216-9

4 5 6 7 8 9 10   2562   25 24 23 22 21

4500844736   A B C D E F G

If you have received these materials as examination copies free of charge, Houghton Mifflin Harcourt Publishing Company retains title to the materials and they may not be resold. Resale of examination copies is strictly prohibited.

Possession of this publication in print format does not entitle users to convert this publication, or any portion of it, into electronic format.

Kadeem is excited.

His dad just bought stamps for Kadeem's collection.

"I can't wait to put them in my album," says Kadeem.

"Animals are my favorites," says Kadeem. "I have 21 on this page. And now I have 8 new ones."

How many animal stamps will he have in all?

Kadeem's stamps on this page are very old.
He takes special care putting 5 more stamps in the album.

How many stamps will there be on this page?

Daniella comes into his room.
"Look," she says. "I bought 3 stamps for you at the post office."
"Great," says Kadeem. "I've already put 17 on this page."

How many will he have when he adds 3 more?

Kadeem's mother brings
7 stamps home from her trip
to California.
Kadeem adds them to his
California page.

How many California stamps does Kadeem have now?

Kadeem closes his album.
Almost all the space in it is filled.
"This is a very, very good day,"
he thinks.

7

Responding

Math Concepts

Kadeem's Album

Draw

Look at page 4. Draw the album page you see. Draw an X wherever you see a stamp.

Tell About

Predict/Infer Look at page 4. Tell how many stamps are on this page of Kadeem's album. Tell how many stamps Kadeem is going to place in the album.

Write

Look at page 4. Write how many stamps Kadeem will have on this page in all.